AIChE Equipment Testing Procedure

TRAY DISTILLATION COLUMNS

A Guide to Performance Evaluation

Second Edition

Prepared by the
Equipment Testing Procedures Committee

Copyright 1987
American Institute of Chemical Engineers
345 East 47 Street, New York, NY 10017

TABLE OF CONTENTS

ISBN 0-8169-0404-9

100.0 PURPOSE AND SCOPE

101.0 *Purpose*

This testing procedure offers methods of conducting and interpreting performance tests on tray distillation columns. Such tests may be made to accumulate reliable data in one or more of the following areas of interest: tray efficiency, capacity limitations, energy consumption, or pressure drop considerations. These data are useful for troubleshooting, for debottlenecking, for new designs, for correlations, and for determination of the operating range of a column as well as the optimum operating conditions. These data may also be required for an "acceptance test."

102.0 *Scope*

Rather than compulsory directions, a collection of techniques is presented to guide the user. Emphasis is placed on principles rather than on specific steps.

This procedure covers continuously operated tray distillation columns. The tray columns include all types of trays, with or without downcomers for liquid. Testing of auxiliary equipment is not included, even though it may limit column capacity. Other established procedures are available for testing auxiliary equipment.

Batch distillation columns are covered in that their performance can be tested best by running them under steady state conditions, either at total reflux or with the overhead returned to the kettle continuously. The principles outlined herein for continuous columns will then apply.

103.0 *Liability*

AIChE and members of the various committees involved make no representation, warranties or guarantees, expressed or implied, as to the application or fitness of the testing procedures suggested here for any specific purpose or use. Company affiliations are shown for information only and do not imply procedure approval by the companies listed. The user ultimately must make his own judgment as to the testing procedures he wishes to utilize for a specific application.

200.0 DEFINITION AND DESCRIPTION OF TERMS

201.0 *Flow Quantities* (Refer to Figure 802.1.)

201.1 *Feed* is the material to be separated, including multiple feed streams.

201.2 *Bottoms* is the high-boiling product leaving the bottom of the column (or the reboiler).

201.3 *Distillate* is the product distilled overhead. It may leave the distillation system as a vapor, liquid, or combination of both.

201.4 *Side-stream product* is a product withdrawn from an intermediate section of the column.

201.5 *Overhead vapor* designates the vapor from the top of the column and includes material to be condensed for reflux. It is the combined distillate and overhead reflux.

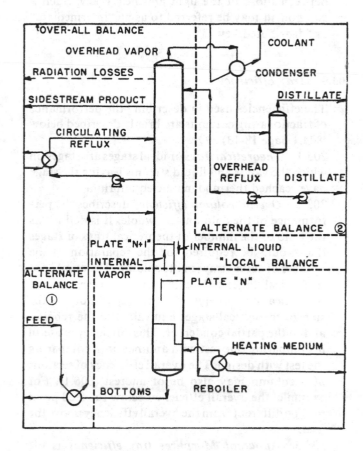

Figure 802.1 Enthalpy balance diagram.

201.6 *Reflux* is used to designate the quantity of liquid returned to the column.

201.6.1 *Overhead (external) reflux* is the quantity of liquid returned to the top tray of the column. Overhead reflux may be subcooled.

201.6.2 *Internal reflux* is the quantity of liquid leaving the top theoretical stage.

201.6.3 *Circulating reflux* is the quantity of liquid withdrawn from the column and returned to the column after being cooled.

201.7 *Throughput* refers to the combined liquid and vapor traffic passing through a cross section of the column.

201.7.1 *Internal liquid* is the calculated quantity of liquid overflowing from tray to tray in the columns.

201.7.2 *Internal vapor* is the calculated quantity of vapor passing from tray to tray in the column.

201.7.3 *Entrainment* is the liquid carried from one tray to another by the vapor stream.

202.0 *Key Components*

The "light key" and "heavy key" components are those two components in a multicomponent mixture between which the "split" is made. At times there may be a component with a boiling temperature between those of the light and heavy key. Such a component may be referred to as an "intermediate" or "distributed key."

203.0 *Tray Efficiency*

The efficiencies used in describing the performance of fractionating columns are briefly described below (804.1 page 18-13).

203.1 *Theoretical trays* or ideal stages are stages on which the vapor and liquid streams leaving the stage have reached thermodynamic equilibrium.

203.2 *Overall column efficiency* describes the performance of the column as a whole. It is defined as the ratio of the number of theoretical trays or stages that would be required for the separation to the number of actual trays in the column. The number of theoretical stages in the column is the total number of theoretical stages required for the separation less the sum of theoretical stages equivalent to the reboiler and to the partial condenser. This efficiency is useful in comparing one test with another or in comparing the test with design. The overall efficiency of sections of a column may also be of interest (606.1). For example, the overall efficiency below the feed point may be different from the overall efficiency above the feed.

203.3 *Apparent Murphree tray efficiency* is the efficiency that could be measured by taking samples around a single tray. It is defined as the actual change in composition accomplished by the tray divided by the change that would occur on a theoretical tray. It accounts for the effects of entrainment, weeping, liquid mixing, maldistribution, etc.

203.4 *Ideal Murphree tray efficiency* describes the performance of a single tray exclusive of the deleterious effects of entrainment, weeping, and liquid backmixing. The ideal Murphree tray efficiency can be predicted from Murphree point efficiency.

203.5 *Murphree point efficiency* is the Murphree efficiency at a single point on the tray.

204.0 *Operating Lines*

These are the material-balance lines on a McCabe-Thiele type of diagram for a binary system (804.2 p. 551). The use of operating lines has been extended to multicomponent systems (804.3).

205.0 *Pinch*

This term describes a local condition within the column under which there is no measurable change in composition of the liquid or vapor components from tray to tray when such lack of change is not due to column malfunctioning (that is, flooding, high entrainment, dry trays). For a binary system a pinch is graphically depicted when an operating line is approaching or intersects the equilibrium curve on a McCabe-Thiele diagram.

206.0 *Maximum Throughput*

206.1 *Maximum hydraulic throughput* is defined as the highest loading at which a column can operate without flooding. Since the loading is affected by both liquid and vapor rates, there are many combinations of these rates which define a maximum-hydraulic-throughput curve.

206.1.1 Flooding describes the condition of the column when the throughput capacity is exceeded. Liquid is being added to a particular tray in the column faster than it will flow from the tray. The column above this point becomes filled with froth and liquid, and liquid is eventually carried out the top of the column. The flooding point can be recognized by the existence of one or all of the situations described in section 502.1

206.2 *Maximum throughput for acceptable separation* is the highest loading at which an acceptable overall column efficiency is obtained. Since loading involves both liquid and vapor, a maximum-throughput-for-acceptable-separation curve can be defined on a plot of liquid vs. vapor rates.

207.0 *Minimum Throughput*

Minimum throughput is the smallest loading at which acceptable performance is obtained. The performance may become unacceptable because of poor separating efficiency due to excessive weeping or liquid/vapor maldistribution, or because of column instability.

208.0 *Operating Section*

An operating section of a column is defined as a portion of a column to which no feed is added, product removed, nor external heat is added or removed.

209.0 *Hardware*

209.1 *Trays* (plates) are baffles used inside columns to effect vapor and liquid contact.

209.2 *Downcomers* are the areas through which the liquid passes from one tray to another.

209.3 *Decks* are the horizontal portions of trays and usually contain the vapor dispersers (i.e., valves, sieve holes, etc.).

300.0 TEST PLANNING

301.0 *Safety*

Any equipment testing must conform to the latest requirements of applicable safety standards. These include, but are not limited to, plant, industry, local, state and federal regulations. It is recommended that all testing be conducted under the supervision of personnel fully experienced in plant and equipment operating practices.

302.0 *Environmental*

The test procedure must conform to the latest requirements of applicable environmental standards which include plant, industry, local, state and federal regulations. Environmental conditions that apply to the equipment in normal operation should also apply during testing.

303.0 *Preliminary Preparation*

The cost of performing plant tests goes far beyond the time and material expended during the actual test run. Careful planning and preparation are essential to maximize the economic and technical benefits of a test. Refer to reference 804.4 for an in-depth discussion of preliminary test preparation.

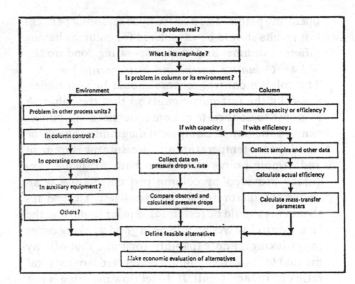

Figure 802.2 Logic diagram for distillation column troubleshooting.

Reprinted from D.B. McLaren and J.C. Upchurch, *Chemical Engineering*, **87**(6), 126 (Aug. 11, 1980).

303.1 *Test Objectives*

Specific objectives should be stated for the plant test. If the test is proposed for plant troubleshooting, a logic diagram similar to Figure 802.2 is useful for defining the nature of the problem and in formulating test objectives (804.5). This phase of test planning should include techniques for determining whether or not the objectives are being met while the test is in progress.

303.2 *Organizational Resources*

The test objectives help determine which organizational resources should be involved in the test. A test team may be formed that would typically include a production engineer (operator interface and experience with equipment involved), a distillation specialist, and a control applications engineer (instrumentation troubleshooting). Plant tests may also involve people from plant maintenance, analytical services, or research and development.

303.3 *Schedule*

Early consideration should be given to the test schedule. Normally the test should be scheduled at a time when unit raw materials are reasonably constant. It should be realized that the test may be conducted under extremely unfavorable conditions, and alternate plans should be provided. Feedstock and product inventories should be compatible with potential increased or decreased production rates resulting from the test. The length of the test will depend on the test objectives, the number of conditions tested, and the time required to reach steady state for each test condition. (See section 503.0.) The number of samples to be collected and types of sample analysis may affect the test schedule since laboratory support is frequently a "bottleneck" in operating plants. The possibility of season-dependent test results should be considered for columns having different summer and winter operating conditions.

303.4 *Column Control and Instrumentation*

The column control scheme should be well understood by the test participants so that the column's response to deliberate or consequential disturbances can be anticipated. Equipment diagrams showing the locations for temperature and pressure measurement and sample points should be made available. Accuracy provided by column instrumentation is an important factor for a successful test. Critical instruments should be recalibrated prior to starting the test. Control valves should be in good working order (nonsticking), and capability for tight shut-off (by manual block valves) should be verified (so that total reflux is, in fact, total). If direct flow metering is not possible on critical streams, plans should be devised for calculating the flows based on known data. Preparations must be made to enable piping, instrument and equipment changes to be effected where necessary to obtain the operating conditions, samples, and data required.

303.5 *Peripheral Equipment*

Potential limitations of peripheral equipment such as reboilers, condensers, pumps, and valves should be considered. Effects of the test on downstream equipment or other processing units must be considered. This is especially important when testing distillation columns that are thermally integrated with other operations.

303.6 *Pretest Calculations*

303.6.1 Simulation of anticipated test conditions is helpful when computer models of the process are available. Flow diagrams showing enthalpy and material balances are instructive for test participants. Frequently, results from simulating a sequence of step changes in operating conditions can be shown graphically. These graphs are useful for comparing actual and anticipated test results while the test is in progress. Another useful pretest calculation is the expected pressure drop for zones of trays and for the entire column. The importance of these calculations is discussed in sections 305.4, 502.1.1, and 605.0. The pretest calculations will require the accumulation of appropriate physical and thermodynamic property data as summarized in Section 307.1.

303.6.2 A useful technique is that of conducting a "dry run" prior to the official test. This pretest, followed by rough processing of the data, enables one to spot shortcomings in data collection; detect factors causing poor closure of mass, component and enthalpy balances; and test the computer simulation (if available) to be used for analyzing test data.

304.0 *Types of Tests*

304.1 *Performance Tests*

Performance tests are conducted to obtain data in one or more of the following specific areas of interest: tray efficiency, capacity limitations, energy consumption, pressure-drop considerations. These data are useful for troubleshooting, for new designs, for correlations, and for determination of the operating range of a column as well as the optimum operating conditions.

304.2 *Acceptance Tests*

An acceptance test is a special type of performance test. Ideally, the conditions under which it is run are those for which the column was designed or guaranteed. The information may be desired only to check that the compositions and production capacity of the distillate and bottoms products are as good as or better than guaranteed and that the utilities consumption is below that specified in the guarantee.

304.2.1 Those responsible for operating the unit should make every effort to achieve design feed rates and compositions for a reliable acceptance test. Frequently, however, the conditions for the test cannot be the same in all respects as those for which the equipment was designed. In this case a test can still be made to check the column, and test results can be extrapolated to the design conditions by techniques based on sound engineering principles. Refer to sections 606.0, 606.1, 606.2, 607.0 and 704.0.

305.0 *Specific Areas of Interest*

305.1 *Tray Efficiencies*

The interest in tray efficiency may be limited to overall column efficiency or may include apparent Murphree tray efficiencies for several or all sections of a column. In conducting efficiency tests, every possible source of error must be minimized. Testing at total reflux minimizes errors in heat input, feed tray composition, and other data required to define operating lines; however, the efficiency of a column operating at total reflux may differ from its efficiency at its design reflux ratio. The choice of total or finite reflux test depends upon the information desired. For determining optimum design conditions and number of trays for new designs, finite reflux ratio tests are recommended.

305.1.1 Murphree tray efficiencies

If the performance characteristics of a particular type of tray are desired, it is necessary to obtain data to calculate the apparent Murphree tray efficiency. The only reliable method of obtaining such data is to take samples from enough trays in a column to establish a complete concentration gradient. Even then the only regions where good tray efficiency data may be obtained are those where appreciable changes in the concentrations of the keys take place from tray to tray.

305.1.2 Overall column efficiency

Results from overall column efficiency tests can be useful in determining optimum feed tray location, optimum reflux ratio, or proper design for new equipment in a similar service.

    a. In many cases better separation of the key components can be effected by feeding to a different tray. Where multiple feed points have been furnished on the column, tests at various feed locations should be made to determine the optimum.

    b. Pinched conditions at the feed tray should be avoided if the test is to measure the true separating ability of the column. A constant temperature zone above or below the feed tray may indicate a pinch at this point. This type of pinch can frequently be overcome by relocating the feed or by increasing the reflux ratio. The amount of pinching that can be tolerated at the top or bottom depends on how accurately the products can be analyzed.

    c. Overall column efficiency tests are valuable for comparison with efficiencies obtained from other columns operating on the same system at the same relative approach to flooding.

305.2 Capacity Limitations

Knowledge of the capacity limitations, both maximum and minimum, of a column is useful when the system is subject to feed rate variations and when future expansion plans or throughput reductions are contemplated.

305.2.1 Maximum hydraulic throughput

At maximum throughput for stable operation the upper limit of vapor-liquid flow capacity of at least one of the trays has been reached and the column begins to flood. It should be realized that the maximum liquid loading depends upon the vapor rate and the maximum vapor loading upon the liquid rate. Furthermore, the vapor and liquid rates usually vary from one section of the column to another. Therefore, several types of capacity tests should be considered if auxiliaries permit:

    a. Determination of maximum feed rate while reflux and reboil rates are increased in proportion.

    b. Determination of maximum reboil rate with reflux to balance at a constant feed rate.

Two additional tests may be useful if a feed preheater is in service:

    a. Determination of the capacity of the section of the column above the feed by increasing the feed preheat (even into the zone of partial feed vaporization) and reflux to balance.

    b. Determination of the capacity of the section of the column below the feed by increasing the reboiler heat and decreasing the feed preheat to compensate so that the reflux rate remains constant.

The detailed procedure for these tests appears in section 502.0.

305.2.2 Maximum throughput for acceptable separation

A drop in separation efficiency to a level where it is no longer acceptable is usually associated with flooding. On other occasions (such as vacuum distillation), excessive entrainment will be the cause of the efficiency drop, much before the column becomes inoperable due to flooding. To find the maximum throughput above which design (or some standard) separation cannot be met, a series of tests is recommended. Tests are run at increasing feed rates with the boilup ratio and reflux ratio held constant. A performance curve can then be plotted from which the maximum throughput for acceptable separation may be determined.

305.2.3 Minimum throughput for acceptable separation

Malperformance at minimum throughput could be the result of liquid/vapor maldistribution, excessive weeping of liquid through the tray bubbling area, leakage through or around trays, or pulsation of vapor flow. In any case, the procedure would be to decrease the reflux, boilup, and feed by the same proportion until the separating action of the column falls off sharply and then to perform the test just above the minimum throughput value so determined. In the case where the minimum throughput is to be that throughput below which design (or some standard) separation cannot be met, a series of tests is recommended at diminishing feed rates (and reflux and boilup in proportion). A performance curve can then be plotted from which the minimum throughput for acceptable separation may be determined. Tests of this type will establish the minimum economical operating rate for an individual column.

305.3 Energy Consumption

A frequent objective of column testing is to determine optimum operating conditions to minimize energy consumption. Two general types of tests are recommended:

    a. Comparison of energy consumption at a fixed separation for various operating conditions. For example, reflux ratio and reboiler duty can be compared for different feed temperatures or different feed trays while maintaining fixed feed composition and product purities.

b. Comparison of energy consumption with reduced product purities. At a fixed feed rate and composition, the reflux ratio and reboiler heat should be gradually reduced. A plot of product purities vs. energy consumption can be developed. The economic advantage of reduced energy consumption can be compared with the economic penalty of reduced product purity.

305.4 *Pressure-Drop Restrictions*

Pressure drop is frequently the most revealing distillation process variable that can be measured. It is important for almost all types of column testing. Measurement of the differential pressure across each operating section is strongly recommended; an overall measurement is essential. Pressure drop measurements are needed when capacity tests are made. (Refer to Section 502.1.) They are invaluable when one is trying to locate and evaluate sections of a column suspected of not operating properly. In addition, pressure drop may be a critical process variable (as in columns handling thermally unstable materials).

306.0 *Data Requirements—Measured*

The data to be collected for any of the tests described above should include all that is necessary to make overall material balances, component balances, and enthalpy balances around the column. Test planning should also include a complete list of all data to be collected. The use of data-logging computers simplifies recording the data. A printout of critical data points at intervals ranging from 1 min to 10 min is a valuable aide in determining when steady state is reached. Some data are best represented by strip-chart recorders (804.5). If data must be recorded by hand, a prepared data form minimizes collection errors, forgetfulness, and collection time (804.4). The number of streams to be sampled, the frequency of sampling, and methods of analysis should be evaluated as a part of test planning.

306.1 *Properties of External Streams, Overall and Component Material Balances, Overall Enthalpy Balances*

306.1.1 The flow rate, temperature, pressure, and composition of the feed stream (or streams), distillate product stream (or streams), bottoms product, and side-stream products must be obtained. In the case of two phase feeds, the temperature and pressure conditions at the last point where the feed is known to be single phase should be obtained. The flow rate, temperature, and composition are required for the reflux liquid stream returning to the top of the column,

and for any circulating reflux in and out of the column and for any special streams such as the solvent in extractive distillation or the entrainer in azeotropic distillation. The pressure and temperatures of the overhead vapor should also be obtained.

306.1.2 To confirm the overall enthalpy balance it is necessary to determine the heat quantities added to or removed from the system by measuring the flow rate, temperatures, and pressure (where vapor exists) of the heating or cooling fluids from exchangers such as reboilers, condensers, feed preheater, etc., passing through the enthalpy envelope. (See Figure 802.1.) These heat quantities, together with the enthalpies calculated for the feed and products, will permit checking the overall enthalpy balance. Radiation and convection losses are usually negligible except in special cases. Vacuum columns are an example since the flowing streams are small compared to the size of the equipment.

306.2 *Internal Temperatures*

Temperature measurements at several points within the column are extremely useful in aiding the calculation of internal flows and in analyzing the performance of the column.

306.2.1 If heat balances are to be hand-calculated, the temperatures of the liquid to and the vapor from any tray where the maximum or minimum flows occur are required in each section with a given tray design. (Refer to section 603.2.) This permits making a heat balance with the envelope intersecting the column at these points so that the internal flows may be calculated.

306.2.2 As discussed in Section 603.2, the calculation of internal flow rate and temperature profiles is much easier (and usually more accurate) when a distillation computer simulation program is available. It is still beneficial to have temperature measurements at several points within the column to be compared with the calculated profiles. However, measured temperatures of the liquid to and vapor from the tray where maximum or minimum flows occur would not be necessary.

306.2.3 In tests where tray efficiencies are to be determined and where a critical analysis of the separation performance is desired, a carefully measured temperature profile of the whole column is advantageous. The temperature points should be strategically placed to cover the zones of maximum temperature change. Alternately, in small fractionators (less than 40 trays), the temperature points should be evenly distributed every five to ten trays. In multicomponent columns where there are moderately large amounts of nonkey components in the feed, the composition of the nonkeys changes rapidly at the feed tray and near the bottom and top of the column; consequently these regions may have large temperature changes.

306.3 *Internal Samples*

Samples of vapor and liquid from within the column are frequently difficult to obtain and should be attempted only if absolutely necessary to achieve the test objectives. Obtaining a representative sample of the vapor leaving a large fractionating tray is unlikely. It is strongly recommended that one rely on liquid samples, obtaining the average vapor composition leaving a tray by calculation. Samples taken above and below the feed tray, preferably from the feed tray and the tray above it, may be desirable for facilitating interpretation of test results.

306.3.1 For accurate tray efficiencies internal samples (preferably liquid) should be withdrawn from a number of trays in the column sufficient to establish a complete concentration gradient across the section in question. (See sections 305.1, 305.1.1, and 404.2.2.)

306.3.2 In tests for maximum and minimum throughputs and overall column efficiency, internal samples re not required since internal compositions can be estimated accurately enough for fluid densities and enthalpies. This point should be checked, however, by a preliminary heat balance.

306.4 *Pressure Profiles*

Pressure-drop measurements are always desirable. (See section 305.4.) Accurate differential meters or manometers are required. (See section 403.4).

307.0 *Data Requirements—Physical Properties*

307.1 *Essential Data*

Essential physical property data must be available for calculating and interpreting the performance data. Essential data of vapor and liquid streams over the range of column conditions include densities, molecular weights, latent heats, and heat capacities (or enthalpy correlations). Heats of solution should be included when this property has an appreciable effect on stream enthalpies. Vapor-liquid equilibrium data are required for graphical tray counting or computerized column simulation. It should be emphasized that accurate VLE data or correlations are essential for meaningful tray efficiency determinations.

307.2 *Auxiliary Data*

Other physical properties useful in developing and testing correlations and for analyzing the performance data include viscosity, diffusivity, and surface tension.

308.0 *Test Procedure Documentation*

The final phase of test planning should be a written document containing all pertinent test planning and information. This document should include a summary of test objectives, manpower requirements, essential physical property data, pretest calculations, data collection procedures, sampling schedules, test sequence, and special safety and environmental considerations. References should be made to the relevant standard sampling procedures, equipment data, and safety and environmental considerations. Finally, a session should be held with the unit operators to fully explain the test and their duties and responsibilities during its operation (804.4).

400.0 METHODS OF MEASUREMENT AND SAMPLING

401.0 *Measurement of Temperatures*

Any reliable thermometer may be used, but thermocouples or resistance thermoelements placed in thermowells are preferred due to their ruggedness and availability in commercial installations. Before use, all temperature-measuring devices should be checked and should be calibrated in the temperature ranges to be used.

401.1 *Accuracy*

The temperatures should be measured with such accuracy that the maximum cumulative error of the heat quantities calculated from temperatures and flow rates is less than 5%. For example, in measuring the heat removal from a condenser by water temperatures, measurements within a fraction of a degree are sometimes required. A relatively accurate temperature difference can be obtained by connecting two thermocouples in series opposing each other. Temperatures of other streams into and out of the column are less critical to the overall heat balance.

401.2 *Errors*

Errors in temperature measurement can still occur in spite of the fact that the sensing element is accurate.

401.2.1 To eliminate errors due to conduction and radiation, recommended techniques should be adhered to (804.6).

401.2.2 Errors due to stratification of a stream may be reduced by installing the thermowell in a location of high fluid turbulence such as downstream from mixing sections or bends. Part of the piping system may serve as a mixing section if radiation losses from the piping are kept small. For liquid temperatures inside the column, thermocouples should be placed in the bottom of the downcomers.

401.2.3 All thermowells should be inspected before a test is made in installations where deposits may occur that would result in erroneous measurements.

401.2.4 In thermosyphon and some other types of reboilers where the outlet line to the column contains two phases, quite different temperature readings will result depending upon the location of the thermocouple in the outlet line, with respect to height. This is due to the continual flashing of the material as the hydrostatic head becomes less in the rising fluid. It is recommended that the thermocouple be placed as near to the column as possible to get a meaningful fractionator temperature. Of course, a thermocouple directly at the reboiler outlet will be useful for observing reboiler performance.

401.2.5 A thermocouple in the feed line to a column requires the same care as in reboiler return lines, due to flashing of the feed as it rises to the feed tray. The preferred location for this thermocouple is near grade level before the feed control valve, where the feed is known to be all liquid. In any case, the thermocouple should be located where the phase condition of the feed can most accurately be known.

402.0   *Measurement of Flow Rates*

The rate of flow may be measured by means of instruments such as orifices, venturi meters, rotameters, displacement meters, vortex meters, or by direct volume or weight measurement. Whenever possible, the instruments should be calibrated in place with the fluid to be measured and the temperature at which it is to be used.

402.1   *Orifice Meters*
Orifice meters have been extensively investigated, and their performance can be accurately predicted for a wide variety of conditions. Details concerning the construction, installation, and calibration of orifices and nozzles can be found in reference 804.7. The procedure outlined on page 5-12 in reference 804.1 may also be followed.

402.1.1   The reading of orifice or venturi-type meters is affected by the density of the flowing material. If the density differs from that for which the meter was calibrated, the reading, if in volumetric units, must be corrected as follows:

volumetric flow rate =

$$(\text{volumetric reading}) \times \frac{\text{design density}}{\text{actual fluid density}}$$

If the meter is also designed to read volumetric flow corrected to a standard temperature and pressure, a further correction is needed if the density differs from the design density of the meter. The combined corrections give:

volumetric flow rate (at standard conditions) =

volumetric reading at standard conditions

$$\times \frac{\text{actual density}}{\text{design density}} \times \frac{\text{design density at standard cond.}}{\text{actual density at standard cond.}}$$

402.1.2   The orifice plate should be checked to verify that it is the proper size and that it is in good condition. Failure to check the orifice plate has ruined results of many test runs.

402.2   *Rotameters*
When rotameters cannot be calibrated in place, generalized charts from manufacturers are usually available. The theory developed by Head (804.8) may be used to convert the calibration from one fluid to that of another as well as for making corrections for variation in temperature and pressure. This procedure is also given on page 5-18 of reference 804.1.

402.3   *Direct Volume or Weight Measurement*
Direct volume or weight measurement should be made wherever practical as a check on other flow measurements. The fluid in tanks or gas holders is weighed or measured volumetrically at specified intervals of time, and a plot of several measurements vs. time is desirable. In situations where there are no steam meters or when it is desirable to check their accuracy, steam condensate can be collected into a large container. When accurately timed, this technique gives an accurate measure of steam flow rate. To minimize condensate flashing and for safety considerations, a condensate cooler should be used.

403.0   *Measurement of Column Pressure Drop*

A differential type of instrument or manometer is preferred. The instrument is connected to the tower through pressure tap lines, which are usually purged or filled with a sealing fluid (section 403.2). Seal pots are also recommended at each tower connection (section 403.3).

403.1   *Instruments*
Manometers are the most convenient instruments to install for temporary measurement of pressure drop. They are portable, are easily installed, and have a wide range of operation. Differential pressure cells are recommended for high-pressure columns and for permanent installations where remote reading is desired. The cell can be connected to the column with or without the use of seal pots.

403.2   *Pressure Taps*
It is necessary to purge or fill the pressure-tap lines with a suitable liquid or gas, if the vapor is condensible at column pressure and atmospheric temperature or if it is corrosive or toxic. As alternatives, pressure-tap lines may be heated above the boiling point of the process fluid, or pressure transmitters may be installed at both the top and bottom of the column. Many differential-pressure transmitters and recorders are available commercially and are useful for continuously recording column pressure drop.

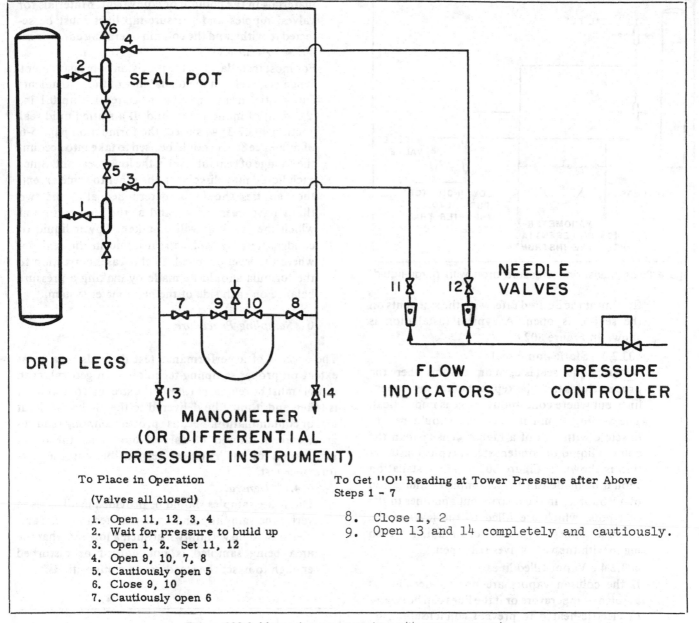

To Place in Operation

(Valves all closed)

1. Open 11, 12, 3, 4
2. Wait for pressure to build up
3. Open 1, 2. Set 11, 12
4. Open 9, 10, 7, 8
5. Cautiously open 5
6. Close 9, 10
7. Cautiously open 6

To Get "O" Reading at Tower Pressure after Above
Steps 1 - 7

8. Close 1, 2
9. Open 13 and 14 completely and cautiously.

Figure 802.3 Measuring pressure drop with purge-gas seal.

403.2.1 Gas purge

A typical installation is shown in Figure 802.3. This technique is the preferred method for keeping lines free from column vapors. The purge flow rate must be steady and low to avoid line pressure drop. A rotameter or bubbler is used to indicate the flow rate, which is adjusted by the needle valves. The flow rates to each line must be equal; otherwise, errors in pressure drop reading can result. A pressure controller is installed to hold the purge-gas pressure constant at a value higher than the column pressure. Before operation, a pressure test of the system is advisable at column pressure with the purge gas. If seal pots (section 403.3) are used, they will allow condensed vapors from the column to drain back to the column and dampen pressure fluctuations.

The reading includes the static head of vapor in the column. In high-pressure columns, the high vapor density causes the static head of vapor to be significant. The static head in the pressure taps can be compensated for by taking the zero reading at the operating pressure or by calculation. Gas purge systems are not recommended for systems in which build-up of noncondensibles in the condenser cannot be tolerated.

403.2.2 Liquid purge

The use of liquid may sometimes be more convenient than a purge gas; the principles for gas purge would also apply for liquid purge. The installation should be made to permit the complete elimination of air bubbles above the liquid in the manometer to the seal pots on the tower. The zero reading is made with purge liquid

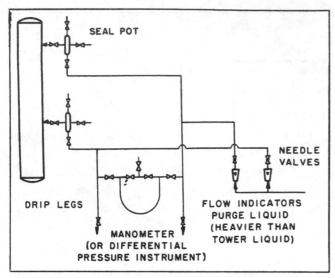

Figure 802.4 Measuring pressure drop with purge-liquid seal.

flowing at the desired rates with the side vents on the seal pots open. A typical installation is shown in Figure 802.4.

403.2.3 Static liquid seal

A static liquid seal is used on columns where the process fluids must be kept out of the instrument lines but where continuous purge is impractical. The sealing liquid in this case should be immiscible with and of a higher density than the column liquid or condensate. A typical installation is shown in Figure 802.5. The installation should be made to permit complete elimination of air bubbles in the instrument and lines to the seal pots, which are filled to the level of the column connections. The zero reading is then made with the vent valves still open.

403.2.4 Vapor-filled lines

If the column vapors are not condensible at ambient temperature or if the lines can be steam- or electric-heated to prevent condensation, no sealing fluid is necessary. The seal pots as shown on Figure 802.3 for the gas purge are still recommended, however, to trap column liquid which may surge into the line.

403.2.5 Pressure-tap line

Flexible high-pressure hose with rapid disconnect pressure fittings is useful for many temporary pressure-drop measurements.

403.3 *Seal pots*

Connections from a column to the pressure-tap lines may be made through seal pots. A suggested size is an 8-in. (about 200-mm) length of 2-in. (50-mm) pipe. The line to the column should be ¾ in. (19 mm) or larger. The design should allow for the connecting line to absorb the total temperature gradient from process fluid to ambient temperature. The other connections should be ¼ in. (6 mm) or larger. The materials of construction should be the same as the

materials in the column or equivalent. Materials for valves, nipples and pressure-tap lines must be selected to withstand the column operating conditions.

403.4 *Accuracy*

For most installations, the precision of measurement depends largely on the readability of the instrument. Differential heads may be measured within 0.1 in. (2-3 mm) of manometer fluid. If a static liquid seal (section 403.2.3) was used, the formula on page 5-6 of reference 804.1 should be used to take into account the change of sealant level in the lower seal pot. Since each liquid may dissolve in the other to some extent, the densities should be determined after the two fluids have been mixed and at the temperature at which the readings will be taken. Tower liquid or condensate may replace seal liquid in the seal pot where the level dropped. In this case, correction to the formula should be made by making a pressure balance on each side of the manometer system.

404.0 *Sampling Procedure*

The success of a performance test depends to a great extent on proper sampling techniques. In general, each sample must be representative of the stream from which it is taken and it must be delivered to the analyst without loss or contamination. Because proper sampling requires great care, it should be carried out under the direct supervision of the engineer responsible for the performance test.

404.1 *General*

Duplicate samples should be provided in all cases to verify the sampling and analytical technique. Samples should be withdrawn at such a low rate that the area being sampled is not starved or disturbed enough to upset the steady-state compositions.

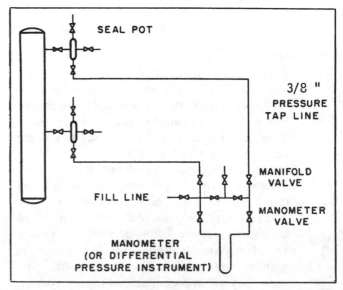

Figure 802.5 Measuring pressure drop with static-liquid seal.

404.2 Selection of Sampling Points

404.2.1 Samples of streams external to the column (feed, reflux, products) should be taken where the stream is all liquid or all vapor. Liquid samples (rather than vapor) give better accuracy. If vapor samples must be used, they should be routed directly to the analyzer via tubing. Locations where stratification or phase separation in the line may exist (for example, at the outlet of a condenser) should be avoided. Points where the fluid is well mixed should be used, for example, following pumps, control valves, but not where vaporization due to flashing exists.

404.2.2 Samples of liquid from within the column should be taken where the liquid is free of vapor. A preferred location is at the base of the downcomer where the liquid is mixed and essentially degassed. Even at this point, however, evidence exists that the liquid may contain entrained vapor, and a settling chamber for the liquid at this location is recommended to provide a vapor disengaging volume and a more representative sample. A liquid sample taken at the base of a downcomer represents the composition of the liquid entering a tray. Liquid from trays without downcomers should be taken from the center by means of a settling chamber.

404.2.3 Samples of vapor from within the column must be free of any entrained liquid. A vapor sampler design that has been used with some success is described in reference 804.9. A modified version of this vapor sampler is described in reference 804.60.

404.3 Sample Connections

To reduce the need for purging large amounts of material, the sample lines should be smaller than ½ in. (13 mm) and preferably ⅛ to ¼ in. (3-6 mm). They should be as short as possible and, for vapor samples, should be free from low sections where liquid may be trapped. A small needle-type valve (sample valve) should be installed downstream from the permanent block valve at the sample outlet. The sample valve is provided in addition to a valve on the sample bomb itself. In cases where a cooling coil is required, it is interposed between the block valve and the sampling valve. A coil of ¼-in. (6-mm) tubing in an open container filled with coolant is usually used. The coil should permit cooling to 80°F (27°C) with 75°F (24°C) water. Additional equipment (pressure gauges, reservoir for liquid used for displacement) is required for bomb sampling. (See section 404.6.)

404.4 Containers

404.4.1 Two types of containers are used depending on the boiling point of the material. (1) Materials with all components boiling above 200°F (93°C) are usually sampled in open containers as liquids. When necessary, the samples should be withdrawn through a cooler. Pyrex bottles or metal cans of corrosion-resistant material are preferred. (2) Materials boiling below 200°F (93°C) are sampled either as a gas or liquid into cylinders or bombs. In addition, the effect of the atmosphere on the sample must be considered; for example, hydroscopic liquids must be sampled in closed containers.

404.4.2 Most companies have standards for construction and testing of containers, but in the absence of these the regulations of the Interstate Commerce Commission are recommended. These regulations also specify sizes and methods of packing and labeling in the event it is necessary to transport samples by common carriers. Cylinders and bombs must have two connections, communicating with opposite ends of the container and closed with valves. For high-pressure bombs, the valves are preferably diaphragm-packed globe valves as used on commercial gas cylinders, with seat side communicating to the container, to minimize danger of leakage. For low-pressure cylinders for gas, small needle valves are suitable. Small glass containers are usually used for low-pressure gas samples if the analytical method does not require large samples.

404.4.3 The size of the container should be large enough to permit at least two accurate analyses.

404.4.4 The containers should be inspected for cleanliness and leaks before being used. Bombs should be pressure-tested if the sample is to be drawn and stored under pressure.

404.5 Sampling of High Boiling Materials

404.5.1 Sampling of a stream that has no component boiling lower than 200°F (93°C) normally can be done in open containers which are immediately stoppered. If a vapor, the sample should be condensed and cooled to at least 80°F (27°C) as it is withdrawn to prevent flashing.

404.5.2 The sample line should first be flushed out to a safe place (or used to rinse the container) with a volume of liquid that is three or four times the dead space of the connections. During sampling, the sample tube should extend into the bottom of the container to prevent splashing and loss of light ends. The container should not be filled entirely, about 10% outage being recommended to allow for expansion, and it should be immediately capped or stoppered.

404.5.3. Caution should be used in handling flammable materials. These fluids usually generate static electricity and where the fluid flows through nonconducting lines or open air, high-voltage charges may build up. It is recommended, therefore, that metal sample containers be grounded to the metal at the sample source.

404.5.4 In cases where extreme care must be used in preventing loss of light material, bombs should be used in place of open containers, and the methods of section 404.6 apply.

404.5.5 The above methods assume that higher than atmospheric pressures are available. Vacuum systems can be sampled by ordinary methods if the sample point (for a liquid) or the sample condenser (for a vapor) is located so that a liquid column sufficient to break the vacuum is available. If this is not practical, a bomb, as described in section 404.4.2, is attached to the sample connection at the lower valve while the upper valve is vented back to a vacuum source by way of a liquid trap. Purging through the sample bomb should be done before sampling.

404.6 *Sampling of Intermediate Boiling Materials* (Approximately –50°F to 200°F or –46°C to 93°C Boiling Point)

These materials are usually sampled in the same physical state as that in which they exist at the sample source. Care must be taken to keep the sample, if vapor, all in the vapor phase. Likewise, if liquid, care must be exercised not to allow any vaporization, since any vapor formed would be trapped in the lines leading to the container. The preferred method of sampling either gas or liquid is to place them into bombs by liquid displacement. An alternative method is by purging. If neither of these procedures can be used because of the nature of the materials being handled, the sample may be taken into an evacuated bomb.

404.6.1 Samples by liquid displacement

The equipment recommended is shown on Figure 802.6A. A suitable liquid for displacement must be chosen so that no transfer of material into the displacement liquid takes place. For hydrocarbons, brine is generally used. The bomb can be filled with the displacement liquid before being connected to the sampling line. This method has the advantage of purging the air from the line above the sample bomb, but this is usually small enough to be negligible. When all

air has been displaced, the sampling line is purged out with the bomb closed off. After the vent has been closed, the procedure varies depending on whether a gas or a liquid is being sampled.

Gas is allowed to flow into the container at such a rate that the pressure is maintained at a value consistent with the design of the sample container or at a rate that will not cause excessive cooling due to rapid expansion. Conditions in the sampling line and container should be such that no condensation of the heavier constituents will take place. If, during purging, "drips" of condensate are observed, no sample should be taken until this condition is corrected. When the desired amount of sample has been collected by measuring the displacement liquid drained from the sample container (or displaced back into reservoir), the valve at the base of the container and the sampling valve are simultaneously closed. The other container valve is then closed and the container may be disconnected.

Liquid is allowed to flow into the container only while pressure is maintained in the container above the vaporization point. If the temperature of the source is above 80°F (27°C), a cooling coil may be necessary to ensure that no vaporization takes place during sampling. All valves are wide open except the vent valve, which is closed, and the lower container valve, which is throttled to maintain flow at the desired rate. The amount of displacement fluid removed is measured, and flow is stopped when the remaining amount of displacement fluid equals the amount of outage required. The container is then closed off from the sampling line and the remaining displacement liquid forced out by the vapor pressure of the material being sampled. The amount of outage required to provide space for liquid expansion is usually 5%, but this may have to be modified for special cases where a significant portion of the lightest component may be transferred into the vapor space. Containers filled with liquid must never be left with both valves closed because of danger of rupture if the temperature increases. The necessity for outage in containers filled with displacement liquid can be avoided by leaving one valve open.

404.6.2 Liquid samples by purging

The equipment required is shown on Figure 802.6B. The sample container is pressured with the sample, and then the compressed vapors and enough liquid to displace the contents of the container and system three times are allowed to escape from the top vent at such a rate that pressure is maintained in the container. Valves are then closed.

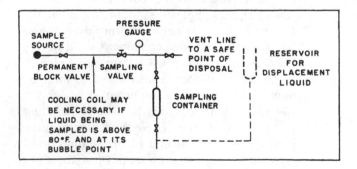

Figure 802.6A Sampling by displacement.

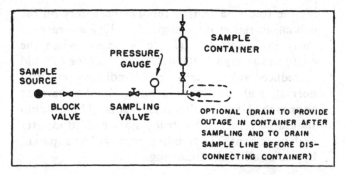

Figure 802.6B Sampling liquids by purging.

Immediately thereafter the necessary outage should be provided. Depending on the materials being handled, it may be advisable to provide a drain for this purpose. For outage requirements and other procedures applicable to pressure containers, see section 404.6.1.

404.6.3 Gas samples by purging

The equipment required is shown on Figure 802.6C. During purging, the sample valve is throttled to prevent exceeding the maximum sample pressure. When at least three times the volume of the sample line and container has been purged through, the sample container valves are closed.

404.6.4 Samples taken into evacuated sample bombs

When this method is used, the sample line is purged and then the bomb is attached to the sample line with suitable fittings. The inlet valve to the sample is opened and then closed, and the bomb removed. Flashing of the liquid inside the bomb may be tolerated as the sample may either be totally condensed before removal from the bomb, or for lighter liquids the entire sample may be removed from the bomb as vapor. Correction should be made for the small amount of air present in the connecting tube between the bomb and the sample line from the column, although this can be eliminated if the bomb and sample line can be evacuated in the field after the two have been connected.

404.7 *Sampling Materials Having Boiling Points Below –50° F (–46° C)*

It is recommended that these materials always be collected as a gas. If a liquid stream is to be sampled, it should be continuously vaporized by a heater placed in the sampling line. Streams having no components that boil above –50° F (–46°C) can be easily vaporized at atmospheric pressure without danger of leaving a small amount of the heavy components behind as liquid. A heating coil may also be necessary for gas samples that are taken from sources having pressures over 300 lb/in.² gauge (2,170 kPa) to prevent cooling, by expansion, below the freezing or condensation point of some of the constituents. The

heater, in any case, should be placed between the block valve and the sampling valve. Otherwise, the equipment is similar to that required for gases (Figure 802.6A or 802.6C).

404.8 *Labeling and Handling the Samples*

404.8.1 The sample should be tagged immediately after being taken and the following information transmitted with the sample:

a. Unit, stream sampled, sampling point

b. Sample number

c. Company and location

d. Date and time

e. Pressure, temperature, and phase of sample at sampling point

f. A note telling analyst how long the sample should be retained before disposing

g. Some idea of the composition of the sample should be given to the analyst so that he can properly prepare his analytical apparatus. If possible, the approximate bubble points of liquid samples should be estimated and reported to the analyst.

h. Identify sample bomb further by name of person conducting test.

i. Weight of bomb after sampling

j. Any special condition that would affect the manner in which the sample should be withdrawn; for example, a statement that the gas was at its dew point (or a liquid at its bubble point) under conditions of subsection e above.

k. Note any hazardous materials or special handling procedures.

404.8.2 Samples should be analyzed as promptly as possible to minimize the change of composition caused by leaks; however, if storage is necessary, liquid samples should be stored in a cool place and vapor samples where condensation will not occur. Storage should be located where leakage will not create fire or toxic hazards. Samples may undergo changes in composition due to chemical reaction, such as polymerization, oxidation, condensation, or decomposition. For this reason, the samples should be analyzed quickly. If not, all possible steps

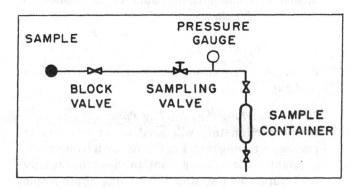

Figure 802.6C Sampling gases by purging.

should be taken to avoid chemical reaction or, at least, to estimate its rate. If the sample is inhibited, this should be noted.

404.8.3 To check for leakage, weighing of bombs is recommended immediately after sampling and just before removal of the analyst's sample.

404.8.4 The analyst should ascertain the condition in which the material exists in the container. In the case of gas samples, heating may be required to vaporize any condensate. For liquid samples under pressure, the liquid should be cooled below the temperature at which it was sampled and the analyst's sample should be withdrawn from the liquid phase in the container. If, during cooling of the liquid sample, the vapor pressure is below that required to remove the analyst's sample, the bomb should be pressurized with an inert gas such as nitrogen or the sample forced out with a displacing fluid such as brine.

500.0 TEST PROCEDURE

501.0 *Preliminary*

An inspection should be made during assembly and/or prior to the start-up of new or revised columns. Where leakage is critical, tests for tray leakage should be made as the column is assembled. In general, it is desirable to know that the internals are assembled properly, that the unit is clean, and that the dimensions, location and orientation of the internals are as specified in the design. A detailed treatment of this subject is discussed in reference 804.61.

All control loops should be operating in a stable manner, i.e., nonoscillating. Loops that are stable can become unstable during test operation at different conditions, particularly as rates are lowered, and could require retuning during the test. Test participants should also recognize circumstances where it may be necessary to modify the existing control strategy or to operate with some control loops off-line in order to obtain the desired measurements. For example, it could be impossible to determine the change of differential pressure as a function of throughput if the column differential is on automatic control.

502.0 *Pretest Procedure for Maximum Hydraulic Throughput*

502.1 *Reaching the Flooding Point*

Many fractionators will need to be operated at maximum throughput. For this reason it is important to establish the flooding point. In any of the capacity tests outlined in section 305.2.1, the approximate capacity (or flooding point) can be found by increas-ing the feed rate and/or reboiler heat duty on the column in rather large steps (5 to 10% increase) at fairly rapid intervals (15 to 30 min). When the column shows symptoms of flooding, the load should be reduced substantially until conditions return to normal, and the flooding point should again be approached in smaller increments. The column should be returned to steady-state condition after each increment. The flooding point will be apparent by one or all of the following:

502.1.1 Differential pressure will increase markedly in the tray zone where flooding takes place. Since only part of the column may be flooded (for example, only the section above the feed tray), the overall column pressure drop may not increase markedly. As a guide, one might expect the pressure drop of each flooded tray to approach the head of clear liquid equal to one-half the tray spacing. Differential pressure measurements across each section of the column are recommended.

502.1.2 Either the bottom flow rate or the overhead product flow rate will drop (or the level in the bottom or reflux drum will drop) as the column holdup increases prior to flood.

502.1.3 An increase in reflux flow rate will not result in an increase in heat input required to maintain the same bottom column temperature. The increased reflux will first accumulate in the column, eventually leave in the overhead, and never reach the bottom of the column.

502.1.4 A rapid increase in temperature may occur below the flooded section as flooding starts because the reduced downflow of liquid from the flooded section leads to that liquid becoming abnormally stripped of its lower boiling components.

502.1.5 A heat balance around the overhead condenser may reveal liquid carryover into the vapor line from the column. This, of course, could be due to high entrainment and therefore is not in itself conclusive of a flooded column.

502.1.6 Other devices such as vapor bleeders for high-pressure columns and sight glasses give visual indication of flooded trays. A description of these devices is given in reference 804.10. Modifications to them may have to be made for reasons of safety or pollution control. A radiation type of detector (such as a gamma-ray detector) can be used without any internal connection to show a high level of liquid or foam on a tray or in the downcomer (references 804.10 and 804.32).

502.1.7 There may be a steep drop in tray efficiency, especially in low-pressure systems, prior to a steep rise in pressure drop if flooding is caused by massive entrainment (jet flooding).

The steep drop in tray efficiency would be apparent by a rapid drop in product purities at a fixed reflux ratio, or by a noticeable rise in column overhead temperature and/or a decrease in bottom temperature.

502.2 *Capacity Test*

After the flooding point has been determined and the column has been unloaded, a test should be made at about 5% below the flooded condition to show that the column is completely operable and to determine steady-state values of the operating conditions.

503.0 *Test Period*

As previously stated (303.3), the length of a test depends on the test objectives, the number of conditions tested, and the time required to reach steady state for each test condition.

503.1 *Establishment of Steady-State Conditions*

The existence of a steady state is established first by observing that column temperatures, pressures, levels, and flow rates have remained constant for a period of time. For columns, in which high-purity products are made, the temperatures in the section that has the highest temperature gradient are the most sensitive and are the best criteria for steady-state conditions. The time required to reach steady state will be greater for columns of large diameter, for those containing a high number of trays and for systems with low relative volatility of the light key. Gilliland (804.11) recommends that this time should be sufficient to displace liquid hold-up in any part of the tower and auxiliaries (including reflux drum and surge in the reboiler circuit) with liquid overflow from two to ten times depending on relative volatility of the light key component, the number of trays, and the desired fractional approach to equilibrium.

503.1.1 In cases in which samples can be analyzed rapidly or continuously, the steady-state condition may be determined by sampling the overhead, bottoms, and feed. At least three successive samples should show essentially constant composition. For columns, in which high-purity products are made, an additional sampling point where the concentration gradient is greatest is useful.

503.1.2 Factors that might prevent steady state are listed in section 701.0. These factors should be reviewed prior to and during the test so that their impact on reaching steady state can be minimized.

503.2 *Length of Test Period*

The length of the test period after reaching steady state depends highly on the test objectives. If the test objective is to ascertain maximum throughput capacity, simultaneous sampling and data collection over a relatively short period of time will give meaningful results. If product or charge tanks are to be gauged to obtain flow rates (or to check flow-rate meters), the test period should be long enough to measure tank displacements within 1 or 2%.

Alternately, the objective may be testing the column in terms of its ability to make a given separation, determining tray efficiency, putting product into storage, or feeding the product to a downstream process. In these cases, only a test of somewhat protracted length can adequately represent the unit's performance. Periodic sampling of the feed and products over a 2-3 day period may be required to average out the effects of small deviations from steady state and errors in laboratory analysis.

600.0 COMPUTATION OF RESULTS

601.0 *Verification of Test Data*

The reliability of the test data must be established before capacity or efficiency calculations can be made. Capacity calculations require constant flow rates and temperatures, but changes in composition that do not affect the enthalpy balance by more than 2% are allowable. Efficiency calculations are more sensitive, and steady-state conditions are required.

602.0 *Material Balance*

The overall material and individual component balances must be made around the column. The allowable amount of error in these material balances depends on the test objectives. Some general guidelines can, however, be given. The overall balance for the column should agree within 3%. If this condition is not met, the data may be unsatisfactory. The balances on the individual distributed components should also agree within 3% and may need to be tighter in order to achieve meaningful efficiency results. The allowable percentage of the discrepancies in the balances on the minor components will depend on the concentrations of those components in the feed. The lower the concentration of a component in the feed for a constant absolute analytical precision, the greater the percentage of discrepancy in the balance of the component is likely to be.

If the balances are unsatisfactory, the data should be examined to make sure that the proper corrections to the instrument readings have been made. The compositions may be checked by means of equilibrium data and the observed temperatures and pressures. An empirical check on the concentrations of the distributed components in the overhead and bottoms streams may be made by plotting, for each component, the ratio of its concentration in the overhead to that in the bottoms vs. its relative volatility at average column conditions. Except for columns that are primarily stripping or rectifying, the plot on log-log paper should give a straight line or a smooth curve approximating a straight line.

The measured flows and compositions should be adjusted to balance the mass flows around the column before any capacity or efficiency calculations are made. In some cases, the flows can be adjusted to improve the component balances. Adjustment of the flows by a least squares method, as described in reference 804.25, should be considered if there are many distributed components. Refer to section 701.1.1 for a list of typical causes of poor material balance closure.

603.0   Enthalpy Balance

603.1   Overall Balance

The consistency of the thermal data on the external streams must be checked before the flows inside the column are calculated. The thermal data can be considered satisfactory if the energy input is within 5% of the energy output. The measured energy input should be taken as the net enthalpy contribution of the entering process streams and the energy added to the reboiler or inter-reboiler (if used). The energy output should be taken as the enthalpy of the leaving process streams, the heat removed by the condenser, and any circulating reflux loops. At times, it may be important to estimate losses to the surroundings due to radiation, convection, or conduction. The same temperature datum should be used to determine the enthalpy of the entering and leaving process streams.

An example of an enthalpy envelope for an entire system is represented in Figure 802.1 by the solid line labeled "overall balance." Several other bases for the enthalpy balance can be used. These are indicated by the dashed lines labeled Alternate 1 and Alternate 2 in Figure 802.1. Consideration should be given to the possibility that some of the process streams passing through the envelope may contain both vapor and liquid phases. If both phases are present, accurate information on their proportions and compositions is required. This information can be used to assist in determining which envelope should be used for the enthalpy balance. It is desirable to have the data for making all three alternate balances. Then a balance around the feed preheater can be used to check the condition of the feed, and a balance around the condenser can be used to determine if entrainment is present. When the distillate is not one pure component, an alternate method for determining entrainment is available. The presence of large quantities of entrainment is indicated when the overhead temperature is lower than that calculated from the composition of the overhead vapor. Refer to section 701.1.2 for a list of typical causes of poor closure of enthalpy balances.

603.2   Internal Flow Rates

Internal flows are determined by making local heat balances (Figure 802.1) or from computer simulations of the column. For maximum (or minimum) capacity tests, the material balance envelope should intersect the column at the plane where the internal flows are the greatest (or least). Preliminary balances may be needed to determine the location of the maximum (or minimum) loading. The determination of operating lines for the calculation of tray efficiencies requires the making of local balances at several locations in the column where major changes occur in the internal flows. If internal samples are not available, the compositions can be estimated with enough accuracy for enthalpy balance purposes. A method for estimating intermediate compositions in an operating column with a multicomponent system is given in reference 804.3.

The task of determining the internal vapor and liquid flows and compositions is much easier if a distillation program is available on a computer. The program can be used to calculate internal flows and external heat duties given the feed, distillate, and reflux rates shown in Figure 802.1. Other combinations of feed, distillate, bottoms, and reflux can be used if certain data are more reliable. Several distillation programs are available through computer services if a program is not available on an in-house computer.

604.0   Column Capacity

The internal flows for the column are calculated from the local heat balances or are available from the column computer calculation (section 603.2). These flows can be used to determine the maximum (or minimum) throughput for acceptable separation and/or the maximum hydraulic load for the column. The column may reach an operating point, before the hydraulic limit, where acceptable specifications cannot be met on the process streams. The internal vapor and liquid rates can be used to explain the loss of separation efficiency at these rates.

The maximum hydraulic limit occurs when any one tray in the column reaches its capacity for handling the vapor or liquid traffic in the column. The flooding of a single tray will cause all trays above that point to fill with liquid. The location of this tray may be evident from observations as described in section 502.1. The vapor and liquid flows for this point should be calculated. If the test run did not define this location, calculations should be made to determine the maximum load point. This point is most easily determined by a computer simulation.

The calculated maximum flows should be used with published correlations to determine if the column is flooding prematurely. Several of the correlations available for various tray types are: bubble cap trays (references 804.13, 804.14, 804.15); sieve trays (references 804.16, 804.54, 804.55, 804.56); and valve trays (references 804.57, 804.58, 804.59). If a comparison with these correlations indicates that the column is flooding prematurely, additional factors that might affect performance should be considered. Some of these factors are tray damage, plugging, improper installation, and foaming. Other possible causes for premature flooding are discussed in section 703.2.1.

605.0 Column Pressure Drop

The internal flows for the column are also useful for calculating the pressure drop to be expected during normal operation. The pressure drop can be calculated from the correlations available in the references given in section 604.0. The comparison of the calculated to the actual pressure drop is very useful in determining if the column is operating properly and how close the system is to the flood point. Many published tray pressure drop correlations do not incorporate the static head of vapor. This omission is usually significant in high-pressure columns. (Refer to section 403.2.1.)

606.0 Overall Column Efficiency

The overall column efficiency is useful for comparing the results of one test with those of another or for comparing test results with the design expectations. For these comparisons to be valid, the same method of calculation, the same equilibrium data, and the same enthalpy data must be used.

606.1 Computer Calculation

A computer calculation of the column is recommended to obtain maximum information from the test data. The calculation assures that the equilibrium conditions and heat and mass balances are satisfied for each theoretical stage. Before any calculations are attempted, it is essential that the consistency of the composition data be checked as described in section 602.0. Most computer programs require the number of theoretical stages as one of the input variables. The usual procedure is to make computer runs with various numbers of theoretical stages in the different sections of the column. A section is a portion of the column over which the vapor and liquid rates are relatively constant. The sections are bounded by the top and bottom, any other intermediate feed or draw trays, or locations of intermediate heat addition or removal. For example, a column with two feeds and one draw has four sections.

The test conditions are used to define the rest of the input data. The number of theoretical stages in each section necessary to yield the product and draw stream compositions that most closely agree with the test results is determined from these runs. The efficiency for each section in the column is probably different. The overall section efficiency is defined by the following equation:

$$\text{Efficiency (section)} = \frac{\text{number of theoretical stages}}{\text{number of actual stages}} \times 100$$

The overall column efficiency can be calculated from the same equation by substituting the total number of stages for the column in the equation.

The interpretation of the results of computer stage-to-stage calculations may be difficult, particularly if there are no obvious key components. A method of finding the column efficiency in such cases is to plot, for each component of the feed, the computer-predicted percentage recovery of that component in the overhead as ordinate vs. the number of theoretical stages as abscissa. The actual percentage recovery in the overhead is then marked on the curve. The corresponding abscissa is the number of theoretical stages of separation given by the column for that component. The indicated numbers of theoretical stages for the several components will not agree exactly. This is due to inaccuracies in the analysis or the physical data or because of differences in the mass transfer rates of the different components. The overall column efficiency for any component can be determined by using this plot and the equation presented above.

606.2 Graphical Calculation

The graphical calculation method (or a computerized version of a graphical method) may be useful if a binary separation is being considered. Two methods have been widely used. The Ponchon-Savarit method (reference 804.2, p. 575) is a rigorous method that allows for deviations from constant molal overflow. The determination of the number of stages may be inaccurate because of small errors in drawing the Ponchon-Savarit diagram. The McCabe-Thiele method (reference 804.2, p. 551) is less sensitive to small graphical errors. It may be used if the operating lines are checked by a heat balance and adjusted as necessary. The Ponchon-Savarit method can be used to determine the operating lines for the McCabe-Thiele diagram.

Multicomponent distillations are studied most easily using a computer program, but a graphical method is available that eliminates the need for stage-to-stage calculations in some sections of the column. The method is given in reference 804.3 and is relatively simple if the liquid compositions on the feed tray and the tray above the feed are obtained experimentally. For the special case where the relative volatilities and molal overflow rates are constant, the equations given in reference 804.17 can be used to compute the number of theoretical stages. An example is the test of a xylene splitter described in the Industrial Data Section of reference 804.12

607.0 *Murphree Tray Efficiency*

Apparent Murphree tray efficiencies can be calculated only when tray samples have been obtained at appropriate points. The efficiencies vary throughout the column, changing with the ratio of the slopes of the equilibrium and operating lines and with composition and velocities in the column (reference 804.13). Since Murphree tray efficiencies are determined at specific velocity, composition, and slope ratio, they are valuable for correlating with fundamental data.

If samples of both the liquid and the vapor to and from the same tray are available, substitution in the formula for the Murphree tray efficiency (reference 804.13, p. 7) is straightforward for binary systems. Usually only liquid samples every five to ten trays are available. A computer calculation can be matched against the data for that section, or the method of reference 804.18 can be used to obtain the efficiency. The graphical method of reference 804.18 requires the use of key components for multicomponent systems. An example of this application is given in reference 804.19. Another article that details the determination of tray efficiencies from operational data is given in reference 804.20.

608.0 *Sample Calculations*

Sample calculations for a methanol/water column are included in section 803.0.

700.0 INTERPRETATION OF RESULTS

This section covers sources of error and gives a check list of possible reasons for the tower not performing as designed, as well as a procedure for converting test results to projected design conditions. Typical questions of test data interpretation include the following: Was the distillation system able to make the required splits, purities, and recovery? If not, why? What might be done to correct this problem? If the design feed composition and amount were not used during the test, what splits would the tower make with the design feed composition and amount at design conditions?

701.0 *Sources of Experimental Error*

Commercial-scale plant tests will normally involve greater errors than those undertaken in research laboratories where equipment and instrumentation are designed specifically for test run data. (References 804.5, 804.21, 804.22, 804.23, 804.24, 804.25, 804.26.)

The following are potential causes of experimental error in commercial plant testing: (a) sampling, (b) analysis, (c) design data, and (d) the presence of unsteady state. These potential errors should be minimized to the extent practical for best test results.

Additional errors in analyzing test data may be introduced by use of incorrect data on the following items: (a) enthalpy, (b) latent heat, (c) vapor pressure, (d) K values, relative volatilities or VLE data, (e) assumption of no heat of mixing, (f) presence of water or other impurities, (g) density of liquids and vapors, (h) viscosity, (i) surface tension, and (j) tray efficiency (leading to inaccurate simulation of the separation).

If there is unsteady state or cycling during the test runs, the results may be meaningless or at best very difficult to interpret. Lack of steady state may be caused by: (a) insufficient time to line out column, (b) varying feed rate and/or compositions (e.g., feed coming directly from a reactor or another column), (c) varying feed enthalpy or percent vaporization, (d) tower cycling due to instrumentation system, poor instrument tuning, or improperly sized valves, (e) improper control points (temperature, pressure), (f) accumulation of water in column or inadvertent refluxing of water into the column, (g) disturbances in heating or cooling media transmitted to the column (e.g., changes in steam pressure, cooling water temperature, etc.), and (h) chemical reaction.

701.1 *Material and Enthalpy Balances*

As discussed in sections 602.0 and 603.0, adequate closure of material and enthalpy balances is of particular importance for proper interpretation of distillation test data.

701.1.1 Inability to close material balances

a. Overall Balance. Typical causes for poor closure of overall material balances are: (a) flowmeter errors, (b) density errors, (c) inventory changes in the tower, exchangers, or reflux drum, and (d) vent losses.

b. Component Balance. In addition to those causes listed above are the following: (a) nonrepresentative samples taken, poor sample technique, (b) incorrect analysis of samples, (c) insufficient number of samples taken to be statistically significant, (d) column not at steady state, inventory changing with time, and (e) heat exchanger leaks.

701.1.2 Inability to close enthalpy balances

Typical causes for poor closure of enthalpy balances are: (a) poor material balance closure, (b) errors in temperature, pressure, composition, or enthalpy of feed or product streams, (c) errors in reflux flow rate, temperature, or composition, (d) entrained liquid from the top tray to the reflux accumulator, (e) errors in flow rate, temperature, pressure or composition of heating or cooling fluids, (f) defective steam traps, (g) inaccurate enthalpy data, or correlations for process, heating or cooling streams, and (h) improper assumptions regarding radiation or convection heat losses or gains.

702.0   *Effects of Experimental Error*

If a computer model of the column separation is available, it is relatively straightforward by case studies to ascertain the effects of experimental error. This can be done by introducing reasonable perturbations to the flow or analytical data from the test run and determining the effect on theoretical tray count or on calculated internal load. The effect of erroneous vapor/liquid equilibrium data can be analyzed in the same way.

703.0   *Possible Reasons for Not Meeting Design Performance*

703.1   *Mechanical*
It may not be possible for the distillation tower systems to reach design capacity and traffic for one or more of the following reasons: (a) inert gases, inadequate condensate removal or undersized condenser, reboiler or heat exchanger, (b) pump problems, (c) insufficient quality or amount of utilities, (d) instrumentation or control problems, (e) insufficient feedstock quality or amount, (f) piping design or installation, (g) mechanical blockages, (h) inability to operate tower at design pressure, (i) equipment damage, (j) errors in equipment location, and (k) equipment fouling (references 804.21, 804.23, 804.27, 804.53).

703.2   *Process*
The major process indications of not meeting design conditions are: (a) tower flooding before reaching design heat duties and internal traffic, (b) actual pressure drop higher than calculated, (c) inability to make design splits, (d) cycling of temperature and inability to reach steady state (804.27), (e) column temperature gradient different from computer simulation or hand calculation, and (f) condenser or reboiler unable to meet design duties.

703.2.1   Premature tower flooding
Distillation towers will flood at tower loads below design for one or more of the following reasons: (a) poor tray design or installation; (b) tray or downcomer blockage or fouling (e.g., lunch pails in downcomer); (c) downcomers too narrow, (d) omission of antijump baffles on trays with high liquid loading and narrow center downcomers, (e) poor internals design or instal-

lation (e.g., reflux or feed distributors, reboiler return, internal baffles, calming zone, weirs, downcomers, seal pans, chimney trays, multipass trays, transition trays, sidestream draw boxes or nozzles, sumps and downcomer trapouts); (f) maldistribution on multipass trays; (g) foaminess; (h) excessive entrainment; (i) vaporization in downcomers; (j) two liquid phases on tray; (k) leaking exchangers or water entering with feeds; (l) insufficient vapor disengagement in downcomers; (m) unsealed or improperly sealed downcomers; (n) excessive downcomer backup; (o) excessive liquid loads; (p) nondesign feed composition or enthalpy; (q) operation at pressure less than design; (r) vapor in liquid outlet lines or vortexing; (s) poor design of instrument nozzles; and (t) damaged trays or internals (references 804.5, 804.22, 804.23, 804.26, 804.27, 804.28, 804.30, 804.31, 804.32, 804.33, 804.34, 804.35, 804.36, 804.37, 804.38, 804.39, 804.40, 804.41, 804.42, 804.43).

The location of the blockage or choke point is usually determined by use of pressure drop measurements, column temperature profiles, or radiation scanning. Cause of localized flooding may be diagnosed most commonly by reviewing tower internals designs and external piping circuits. The possibility of leaking exchangers may be diagnosed by appropriate pressure, vacuum, or leak tests. Entrainment may be checked by injection of dyes or tracers into the feed, reflux, or reboiler return.

The possibility of two liquid phases (water) is normally checked by hand calculations on suspected trays, using the best VLE and LLE data available together with computer simulations of the tower at test run conditions.

Foaming can be checked by small-scale tests using real tower liquids over the operating temperature and pressure range, and bubbling gas through the liquid or suddenly depressuring the boiling liquid. If foaming is noted in the small test, it may be solved normally by continuous injections of an effective defoaming compound.

Maldistribution is especially possible on multipass tray designs. It is most easily checked by tower temperature gradients taken circumferentially around the tower or by radiation scanning.

703.2.2   Higher than design pressure drop
The reasons for high pressure drop are the same as those for premature flooding above. High pressure drop *per se* is rarely a critical problem. Exceptions are separations with temperature-sensitive bottoms products and situations where the high base pressure causes an insufficient temperature driving force in the reboiler.

703.2.3 Inability to make design splits (apparent low tray efficiency)

The inability to make the design splits can usually be attributed to one or more of the following causes: (a) incorrect VLE data used in design; (b) sampling not representative; (c) too few samples to be statistically significant; (d) incorrect sample analysis; (e) incorrect enthalpy or physical properties used in design; (f) incorrect assumption of tray efficiency; (g) poor tray and internals design, fabrication, or installation; (h) damaged, corroded, or fouled trays or internals; (i) unsealed downcomers or inadequate seal pans; (j) maldistribution of vapor or liquid; (k) nonlevel trays or weirs or downcomer seals; (l) excessive tray leakage or weeping; (m) non-design feedstock, composition or enthalpy; (n) underloading or trays; (o) steady state not reached or tower cycling; (p) excessive entrainment; (q) nonoptimum feed tray; (r) vapor, bypassing of tray liquid; (s) poor column control; (t) problems with upstream or downstream equipment; (u) unsuitable design of condenser, reboiler, and other auxiliaries and control valves; (v) incorrect calibration of instruments; (w) incorrect tower operating conditions; (x) incorrect utilities or heating and cooling condition; (y) incorrect reflux temperatures; (z) incorrect overhead draw rate; (za) incorrect instrument installation, leading to incorrect readings or control; (zb) excessive heat leaks; (zc) blockages or malfunctions in control valves or instrument taps or orifices, pressure drop instruments, thermocouples, pressure tap level controls; (zd) air leaks in vacuum towers; (ze) incorrect piping; (zf) exchanger leaks; (zg) contamination of the column feed (e.g., water); (zh) azeotropes; (zi) foaming; (zj) calculation procedure for number of trays incorrect or critical assumptions not valid; (zk) excessive liquid recirculation on trays; (zl) excessive shearing of tray liquid to form "fog" on trays (this occurs at low liquid rates combined with high gas rate—liquid is not able to flow down the downcomers); (zm) use of tray types which are too rate sensitive, particularly where tray traffic varies widely from top to bottom of the column (e.g., demethanizers, depropanizers, extractive distillation, or vacuum columns); (zn) two liquid phases on trays; (zo) chemical reaction in the column; and (zp) leakage of spare feed point block valve (references 804.5, 804.21, 804.22, 804.23, 804.24, 804.26, 804.28, 804.29, 804.33, 804.34, 804.35, 804.36, 804.37, 804.43, 804.44, 804.45, 804.46, 804.47, 804.48, 804.49, 804.50, 804.51, 804.52, 804.53).

A procedure for diagnosing the reasons for column malperformance must start with the questions: Is the problem real? What is the magnitude? A logic diagram for distillation column troubleshooting (Figure 802.2) can be extremely useful. It is usually better to roughly find the problems and develop economical short-term and longer-term rapid solutions than to spend weeks achieving extreme accuracy in defining the tray efficiency found in the test. Comparison of the deficient tower with successful operating towers is also frequently used.

703.2.4 Cycling of temperatures and inability to reach steady state

Typical reasons for cycling of tower temperature or pressure are: (a) incorrect instrumentation system—design or installation, (b) unstable reboiler circuits, (c) unstable trays, (d) heat exchanger leaks, (e) unsteady condensate removal, (f) erroneous instrumentation readings or calibration, (g) two liquid phases on trays (804.37), (h) control valves too large or too small, (i) controller malfunction, and (j) variation in feed rate or composition.

703.2.5 Column temperature gradient different from computer simulation

The major causes for temperature gradients not matching computer simulation are: (a) incorrect overhead draw rate, (b) tray leakage, (c) low tray efficiency, (d) liquid or vapor maldistribution, (e) incorrect VLE, vapor pressure, or enthalpy data (correlations), (f) partial flooding—usually indicated by zones with little or no temperature change, (g) excessive entrainment, (h) water in feed or reflux, (i) leaking exchangers, (j) incomplete mixing of the liquids entering the feed tray, (d) column not at steady state, (l) poor internal distributor design for sidedraws, reflux, feed, or reboiler piping, (m) unsealed downcomers with vapor going up downcomers, and (n) incorrect thermocouple/thermowell installation or calibration (references 804.24, 804.27, 804.28, 804.36, 804.43, 804.50).

704.0 *Application of Test Results to Design Conditions*

Frequently the conditions for the test cannot be the same in all respects as those for which the equipment was designed. If conditions reasonably close to design can be achieved, a test can still be made to check the column, and test results can be extrapolated to the design conditions by techniques based on sound engineering principles. In general, the procedure would be as follows:

a. Using the design VLE and enthalpy data and taking into account the experimental uncertainty, calculate the maximum and minimum numbers of theoretical stages in the rectifying and stripping sections under conditions of the test.

b. Assume that the same number of theoretical stages will be developed under design conditions (i.e., that the tray efficiencies will be the same at design conditions as they were for the test conditions).

c. Using these numbers of theoretical stages, calculate by computer simulation or by hand the best and worst splits of the key components under design conditions.

This technique may be applicable particularly for interpreting results for an acceptance test, section 304.2

800.0 APPENDIX

801.0 Notation

AA	= active area, ft^2 (m^2)
A_H	= hole area, ft^2 (m^2)
CAF	= capacity factor
Delta P dry	= dry tray pressure drop, in. of liquid (mm Hg)
Delta P total	= total pressure drop, in. of liquid (mm Hg)
FPL	= flow path length, in. (m)
H_w	= weir height, in. (mm)
L_{wi}	= weir length, in. (m)
Q	= heat transferred, Btu/h (kW)
Q_L	= volumetric liquid flow, gpm (m^3/h)
V_H	= hole velocity, ft/s (m/s)
V_{load}	= vapor load for any tray in the section ft^3/s (m^3/s)

803.0 Sample Calculations

The data used in this example are for illustration purposes only and should not be used for design.

The primary separation in the column is methanol/water. All of the trace components were ignored to simplify the calculations.

803.1 English Units

The sample calculations will be shown first in English units.

803.1.1 Material balance

Test Data:

Feed	40,000	lb/h
	158	°F
	30	psig
	71.43	wt. % methanol
	28.57	wt. % water
Distillate	28,600	lb/h
	122	°F
	80	ppm water
Bottoms	11,900	lb/h
	226	°F
	2,300	ppm methanol
Reboiler	29,800	lb/h steam

	20	psig, chest
	259	°F, steam chest
	287	°F, steam header
	40	psig, steam header
Condenser	3,100	gpm water
	86	°F inlet
	103	°F outlet

Column top vapor is 150°F

Reflux ratio is 1.0 (L/D, Reflux/Distillate)

Overall Material Balance:

$$\% \text{ error} = \frac{\text{flow(in)} - \text{flow(out)}}{\text{flow(in)}} \times 100$$

$$= \frac{40,000 - (28,600 + 11,900)}{40,000} \times 100$$

$$= -1.25$$

This is acceptable.

Component Balance:

Methanol:

$$\% \text{ error} = \frac{0.7143(40,000) - [0.99992(28,600) + 0.0023(11,900)]}{0.7143(40,000)} \times 100$$

$$= -0.19$$

Water:

$$\% \text{ error} = \frac{0.2857(40,000) - [0.00008(28,600) + 0.9977(11,900)]}{0.2857(40,000)} \times 100$$

$$= -3.91$$

This is acceptable for methanol, but marginal for water. This is an indication that bottoms flow is probably least accurate. Use feed and tops in computer calculations.

803.1.2 Enthalpy balance

Overall Balance:

Datum condition is liquid at 122°F

Heat in = $Q_{reboiler} + Q_{feed}$

= 29,800(1,175.9 − 228.0)

+ 0.7143(40,000)(0.69)(158 − 122)

+ 0.2857(40,000)(1.0)(158 − 122)

= 29.37 × 10^6 Btu/h

Heat out = $Q_{condenser} + Q_{tops} + Q_{bottoms}$

+ Q_{losses} Neglect

= 3,100(1/7.48)(62.0)(60)(1.0)(103 − 85)

+ 28,600(0.67)(122 − 122)

+ 11,900(1.0)(226 − 122)

= 29.98 × 10^6 Btu/h

$$\% \text{ error} = \frac{Q_{in} - Q_{out}}{Q_{in}} \times 100$$

$$= \frac{29.37 \times 10^6 - 28.98 \times 10^6}{29.37 \times 10^6} \times 100$$

$$= 1.33$$

Local Balance:
Alternate 2—Overhead Loop
Datum is liquid at 122°F

Heat in $= Q_{vapor}$
$= 2.0(28,600)(483.6)$
$+ 2.0(28,600)(0.35)(150 - 122)$
$= 28.22 \times 10^6$ Btu/h

Heat out $= Q_{cond} + Q_{tops}^{Q} + Q_{reflux}^{Q}$
$= 3,100(1/7.48)(62.0)(60)(1.0)(103 - 85)$
$= 27.75 \times 10^6$ Btu/h

% error $= \dfrac{Q_{in} - Q_{out}}{Q_{in}} \times 100$

$= \dfrac{28.22 \times 10^6 - 27.75 \times 10^6}{28.22 \times 10^6} \times 100$

$= 1.67$

The overall and local enthalpy balances are acceptable. The feed is subcooled based on equilibrium calculations.

803.1.3 Column capacity

The column flooded at rates slightly above those given in sections 803.1.1 and 803.1.2. These rates are the last stable data and will be used as the basis for a computer simulation of the column. The feed and distillate flow rates along with the reflux ratio will be specified. The number of theoretical stages above and below the feed was varied to match the observed concentrations at top and bottom. The following table summarizes the computer calculation. A column with 20 theoretical stages most closely matched plant data.

Theo. Stage No.	Temp. °F	Pressure psia	Flows, lb/h Vapor	Liquid	Density, lb/ft³ Vapor	Liquid	Methanol Conc. Liquid Wt. %
Cond.	122	—	—	57,158	—	49.19	99.9926
20	150.4	15.43	57,158	29,732	0.0761	49.10	99.9699
19	151.0	15.65	58,306	29,712	0.0771	49.10	99.9226
18	151.7	15.86	58,291	29,677	0.0780	49.11	99.8248
17	152.5	16.08	58,256	29,605	0.0790	49.13	99.6236
16	153.4	16.30	58,184	29,460	0.0799	49.16	99.2135
15	154.5	16.51	58,039	29,172	0.0807	49.24	98.3918
14	156.1	16.73	57,752	28,633	0.0812	49.38	96.7970
13	158.5	16.95	57,212	27,705	0.0815	49.66	93.8780
12	162.2	17.16	56,284	26,315	0.0811	50.11	89.0504
11	167.0	17.38	54,894	24,601	0.0801	50.76	82.2276
10	172.5	17.60	53,180	64,858	0.0787	51.53	74.3946
9	175.2	17.81	53,401	63,409	0.0784	51.86	72.0211
8	180.5	18.03	51,592	59,419	0.0767	52.61	63.5627
7	189.5	18.25	47,961	53,970	0.0724	53.99	50.1816
6	201.3	18.46	42,510	48,408	0.0653	55.84	32.9421
5	211.9	18.68	36,947	44,505	0.0578	57.52	17.8821
4	218.9	18.90	33,043	42,409	0.0525	58.60	8.4506
3	222.9	19.11	30,947	41,443	0.0497	59.14	3.6873
2	225.0	19.33	29,980	41,033	0.0487	59.38	1.5376
1	226.2	19.55	29,571	40,872	0.0485	59.48	0.6159
Reboil.	227.1	19.76	29,407	11,464	0.0487	59.52	0.2292

The points where column loads are highest are stages 10 (liquid) and 19 (vapor). A flood calculation for each of these points should be carried out, and the one closer to the flood point should be determined. This calculation showed that stage 19 is closer to the flood point and is therefore likely to flood first. For simplicity, the flooding calculation is shown only for theoretical stage 19.

Column Data:

30 cross-flow valve trays, feed tray 17	
10-gauge carbon steel decks	
258 14-gauge stainless steel V-1 valves	
Column diameter, ID	66.0 in.
Column area	23.76 ft²
Active area	20.48 ft²
Downcomer area	1.64 ft²
Slot area	3.29 ft²
Flow path length	50.0 in.
Weir length	43.1 in.
Outlet weir height	2.0 in.
Tray spacing	24.0 in.

Calculate percent flood using reference 804.57 and flows for theoretical stage 19.

$V_{load} = 58,306(1/0.0771)(1/3600)[0.0771/$
$(49.10 - 0.0771)]^{0.5}$
$= 8.33$ ft³/s

$Q_L = 29,732(1/49.10)(7.48)(1/60)$
$= 75.49$ gpm

From Figure 5b (Reference 804.57)

$CAF_o = CAF = 0.426$

Equation 13, % flood

$= \dfrac{V_{load} + Q_L (FPL/13,000)}{AA (CAF)} \times 100$

$= \dfrac{8.33 + 75.49(50/13,000)}{(20.48)(0.426)} \times 100$

$= 98.8$

For downcomer:

$VD_{DSG} = 250$ gpm/ft² (From Figure 4 of Reference 804.57)

Downcomer load

$= \dfrac{75.49}{(1.64)(250)} \times 100$

$= 18.4\%$

These calculations indicate that the column is actually flooding in the top section of the column at rates very close to the predicted flooding rates.

803.1.4 Pressure drop

The tray pressure drop should be calculated at several points for comparison. Reference 804.57 is used.

Theoretical Stage 19

$A_H = 258/78.5 = 3.29$ ft² (Eq. 19)

$V_H = 58,306(1/0.0771)(1/3,600)(1/3.29)$
$= 63.85$ ft/s

Equation 18a

$$\Delta P_{dry} = 1.35(0.074)(510/49.10)$$
$$+ 0.2(63.85)^2(0.0771/49.10)$$
$$= 2.32 \text{ in. of liquid}$$

Equation 18b

$$\Delta P_{dry} = 0.82(63.85)^2(0.0771/49.10)$$
$$= 5.25 \text{ in. of liquid}$$

As explained in reference 804.57, the larger value from Eq. 18a or 18b is used for calculating total pressure drop in Eq. 20.

Equation 20

$$\Delta P_{total} = \Delta P_{dry} + 0.4(Q_L/L_{wi})^{2/3}$$
$$+ 0.4\,H_w$$
$$= 5.25 + 0.4(75.49/43.1)^{2/3}$$
$$+ 0.4(2.0)$$
$$= 6.63 \text{ in. } (49.10/62.32)$$
$$= 5.22 \text{ in. water}$$

Additional pressure drop points:

Theoretical Stage	Delta P (in. Water)
19	5.22
17	5.13
15	5.01
13	4.85
11	4.60
9	4.85
7	4.38
5	3.59
3	3.18
1	3.06
Average	4.39

Calculated column pressure drop is 30(4.39) = 132 in. water. Measured column pressure drop was 120 in. water.

803.1.5 Column efficiency

The column efficiency can be calculated from the previous computer calculation.

$$\text{Overall column efficiency} = \frac{\text{no. theo. stages}}{\text{no. actual trays}} \times 100$$
$$= \frac{20}{30} \times 100$$
$$= 66.7\%$$

The individual section efficiencies may also be calculated:

$$\text{Top section efficiency} = \frac{10}{13} = 100$$
$$= 76.9\%$$

$$\text{Bottom section efficiency} = \frac{10}{17} = 100$$
$$= 58.8\%$$

803.2 SI Units

The previous example will be repeated with SI units.

803.2.1 Material balance

Test Data:

Feed	18,144	kg/h
	70.2	°C
	308	kPa
	71.43	wt. % methanol
	28.57	wt. % water
Distillate	12,972	kg/h
	50.1	°C
	80	ppm water
Bottoms	5,398	kg/h
	108	°C
	2,300	ppm methanol
Reboiler	13,517	kg/h steam
	239	kPa, chest
	126	°C, steam chest
	142	°C, steam header
	377	kPa, steam header
Condenser	704	m³/h water
	29.6	°C inlet
	39.6	°C outlet

Column top vapor is 65.7°C

Reflux ratio is 1.0 (L/D, Reflux/Distillate)

Overall Material Balance:

$$\% \text{ error} = \frac{\text{flow(in)} - \text{flow(out)}}{\text{flow(in)}} \times 100$$
$$= \frac{18,144 - (12,972 + 5,398)}{18,144} \times 100$$
$$= -1.25$$

This is acceptable.

Component Balance:

Methanol:

$$\% \text{ error} = \frac{\begin{array}{c}0.7143(18,144)- \\ [0.99992(12,972)+ \\ 0.0023(5,398)]\end{array}}{0.7143(18,144)} \times 100$$
$$= -0.19$$

Water:

$$\% \text{ error} = \frac{\begin{array}{c}0.2857(18,144)- \\ [0.00008(12,972)+ \\ 0.9977(5,398)]\end{array}}{0.2857(18,144)} \times 100$$
$$= -3.91$$

This is acceptable for methanol, but marginal for water. This is an indication that bottoms flow is probably least accurate. Use feed and tops in computer calculations.

803.2.2 Enthalpy balance

Overall Balance:

Datum condition is liquid at 50.1°C

Heat in

$$= Q_{reboiler} + Q_{feed}$$
$$= 13,517(0.612) + (0.7143)(18,144)(0.000080)(70.2-50.1)$$
$$+ 0.2857(18,144)(0.00116)(70.2 - 50.1)$$
$$= 8,603 \text{ kW}$$

Heat out $= Q_{condenser} + Q_{tops} + Q_{bottoms}$
$\qquad + Q_{losses} Neglect$
$\qquad = (704)(993)(0.001162)(39.6 - 29.6)$
$\qquad + 12,972(0.000778)(50.1 - 50.1)$
$\qquad + 5,398(0.00116)(108 - 50.1)$
$\qquad = 8,485$ kW

$\% \text{ error} = \dfrac{Q_{in} - Q_{out}}{Q_{in}} \times 100$

$\qquad = \dfrac{8,603 - 8,485}{8,603} \times 100$

$\qquad = 1.37$

Local Balance:
Alternate 2—Overhead Loop
Datum is liquid at 50.1°C

Heat in $= Q_{vapor}$
$\qquad = 2.0(12,972)(0.3123)$
$\qquad + 2.0(12,972)(0.000407)(65.7 - 50.1)$
$\qquad = 8,267$ kW

Heat out $= Q_{cond} + Q_{tops} + Q_{reflux}$
$\qquad = 704(993)(0.001162)(39.6 - 29.6)$
$\qquad = 8,123$ kW

$\% \text{ error} = \dfrac{Q_{in} - Q_{out}}{Q_{in}} \times 100$

$\qquad = \dfrac{8,267 - 8,123}{8,267} \times 100$

$\qquad = 1.74$

The overall and local enthalpy balances are acceptable. The feed is subcooled based on equilibrium calculations.

803.2.3 *Column capacity*

The column flooded at rates slightly above those given in sections 803.2.1 and 803.2.2. These rates are the last stable data and will be used as the basis for a computer simulation of the column. The feed and distillate flow rates along with the reflux ratio will be specified. The number of theoretical stages above and below the feed was varied to match the observed concentrations at top and bottom. The following table summarizes the computer calculation. A column with 20 theoretical stages most closely matched plant data.

The points where column loads are highest are stages 10 (liquid) and 19 (vapor). A flood calculation for each of these points should be carried out, and the one closer to the flood point should be determined. This calculation showed that stage 19 is closer to the flood point and is therefore likely to flood first. For simplicity, the flooding calculation is shown only for theoretical stage 19.

Theo. Stage No.	Temp. °C	Pressure kPa	Flows, kg/h Vapor	Liquid	Density, kg/m³ Vapor	Liquid	Methanol Conc. Liquid Wt. %
Cond.	50.17	—	—	25,926	—	787.9	99.9926
20	65.96	106.4	25,926	13,486	1,219	786.5	99.9699
19	66.28	107.9	26,447	13,477	1.235	786.5	99.9226
18	66.67	109.4	26,440	13,461	1.249	786.6	99.8248
17	67.12	110.9	26,424	13,429	1.265	787.0	99.6236
16	67.62	112.4	26,392	13,363	1.280	787.4	99.2135
15	68.23	113.8	26,326	13,232	1.293	788.7	98.3918
14	69.12	115.4	26,196	12,988	1.301	791.0	96.7970
13	70.45	116.9	25,951	12,567	1.305	795.5	93.8780
12	72.51	118.3	25,530	11,936	1.299	802.7	89.0504
11	75.17	119.8	24,899	11,159	1.283	813.1	82.2276
10	78.23	121.3	24,122	29,419	1.261	825.4	74.3956
9	79.73	122.8	24,222	28,598	1.256	830.7	72.0211
8	82.67	124.3	23,402	26,952	1.229	842.7	63.5627
7	87.67	125.8	21,755	24,480	1.160	864.8	50.1816
6	94.23	127.3	19,282	21,957	1.046	894.4	32.9421
5	100.12	128.8	16,759	20,187	0.926	921.4	17.8821
4	104.01	130.3	14,988	19,236	0.841	938.7	8.4506
3	106.23	131.8	14,037	18,798	0.796	947.3	3.6873
2	107.40	133.3	13,599	18,613	0.780	951.1	1.5376
1	108.06	134.8	13,413	18,539	0.777	952.8	0.6159
Reboil.	108.56	136.2	13,339	5,200	0.780	953.4	0.2292

Column Data:
30 cross-flow valve trays, feed tray 17
3.4-mm-thick carbon steel decks
258 14-gauge stainless steel V-1 valves

Column diameter, ID	1.676	m
Column area	2.207	m²
Active area	1.902	m²
Downcomer area	0.152	m²
Slot area	0.306	m²
Flow path length	1.27	m
Weir length	1.09	m
Outlet weir height	50	mm
Tray spacing	0.610	m

The precent flood correlation, Eq. 13 of reference 804.57, is empirical in nature and is not readily converted to SI units. As shown in section 803.1.3, the calculated percent flood for theoretical stage 19 is 98.8%.

803.2.4 *Pressure drop*

The tray pressure drop should be calculated at several points for comparison. The correlations used are given in refence 804.57. These correlations are empirical in nature and are not readily converted to SI units. The calculation procedure is the same as shown in section 803.1.4. The results of the calculation are summarized in the following table.

Additional pressure drop points:

Theoretical Stage	Delta P (mm Hg)
19	9.75
17	9.58
15	9.36
13	9.06
11	8.59
9	9.06
7	8.18
5	6.71
3	5.94
1	5.72
Average	8.20

Calculated column pressure drop is 30(8.20) = 246 mm Hg. Measured column pressure drop was 224 mm Hg.

803.2.5 *Column efficiency*

The column efficiency can be calculated from the previous computer calculation.

$$\text{Overall column efficiency} = \frac{\text{no. theo. stages}}{\text{no. actual trays}} \times 100$$

$$= \frac{20}{30} \times 100$$

$$= 66.7\%$$

The individual section efficiencies may also be calculated:

$$\text{Top section efficiency} = \frac{10}{13} \times 100$$

$$= 76.9\%$$

$$\text{Bottom section efficiency} = \frac{10}{17} \times 100$$

$$= 58.8\%$$

804.0 *References*

804.1   Perry, R.H., and C.H. Chilton, "Chemical Engineers' Handbook" 5th Ed., McGraw-Hill, New York (1973).

804.2   McCabe, W.L., and J.C. Smith, "Unit Operations of Chemical Engineering," McGraw-Hill, New York (1976).

804.3   Hengstebeck, R.J., *Petrol. Eng.*, **29**, C-6 (1957).

804.4   Shaw, R.J., J.A. Sykes, and R.W. Ormsby, *Chem. Eng.*, **87**(16)), 126(Aug. 11, 1980).

804.5   McLaren, D.B., and J.C. Upchurch, *Chem. Eng.*, **77**(12), 139(June 1, 1970).

804.6   "AIChE Standard Testing Procedure for Heat Exchangers: Section I, Sensible Heat Transfer in Shell and Tube-Type Equipment."

804.7   ASME Research committee on Fluid Meters Report, "Fluid Meters—Their Theory and Applications," 6th Ed. (1971).

804.8   Head, V.P., *Trans. ASME*, **76**, 851(1954).

804.9   Kastenek, F., and G. Standard, *Separation Sci.*, **2**(4), 439(1967).

804.10   Kelley, R.E., T.W. Pickel, and G.W. Wilson, *Petrol. Refiner*, **34**(1), 110(1955); **34**(2), 159(1955).

804.11   Robinson, C.S., and E.R. Gilliland, "Elements of Fractional Distillation," 4th Ed., p. 476, McGraw-Hill, New York (1950).

804.12   AIChE Research Committee, "Final Annual Report, University of Delaware," AIChE, New York (1958).

804.13   Distillation Subcommittee of the Research Committee, "Bubble Tray Design Manual," AIChE, New York (1958).

804.14   Bolles, W.L., "Optimum Bubble Cap Tray Design," Fritz W. Glitsch & Sons, Inc., Bulletin No. 156-2, Dallas, TX.

804.15   Bolles, W.L., "Tray Hydraulics—Bubble Cap Trays," Ch. 14, Smith, B.D., "Design of Equilibrium Stage Processes," p. 474, McGraw-Hill, New York (1963).

804.16   Fair, J.R., "Tray Hydraulics-Perforated Trays," Ch. 15, Smith, B.D., "Design of Equilibrium Stage Processes," p. 539, McGraw-Hill, New York (1963).

804.17   Underwood, A.J.V., *Chem. Eng. Prog.*, **44**(8), 603 (1948); **45**(10), 609(1949).

804.18   Baker, T., and J.S. Stockhardt, *Ind. Eng. Chem.*, **22**(4), 376(1930).

804.19   Gerster, J.A., T. Mizushina, T.N. Marks, and A.W. Catanach, *AIChE J.*, **1**(4), 536 (1955).

804.20   Taylor, D.L., P. Davis, and C.D. Holland, *AIChE J.*, **10**(6), 864 (1964); **11**(4), 678 (1965).

804.21   Custer, R.S., *Chem. Eng. Prog.*, **61**(9), 86(1965).

804.22   Berg, C., and I.J. James, Jr., *Chem. Eng. Prog.*, **44**(4), 307(1948).

804.23   Shah, G.C., *Chem. Eng.*, **85**(17), 70 (July 31, 1978).

804.24   Martin, H.W., *Chem. Eng. Prog.*, **60**(10), 50(1964).

804.25   Gelus, E., S. Marple, Jr., and E. Manning, Jr., *Chem. Eng. Prog*, **45**(10), 602(1949).

804.26   Kister, H.Z., *Chem. Eng.*, **88**(7), 97(Apr. 6, 1981).

804.27   Hausch, D.C., *Chem. Eng. Prog.*, **60**(10), 55(1964).

804.28   Andersen, A.E., and J.C. Jubin, *Chem. Eng. Prog.*, **60**(10), 60(1964).

804.29   Bolles, W.L., *Chem. Eng. Prog.*, **63**(9), 48(1967).

804.30   Eagle, R.S., *Chem. Eng. Prog.*, **60**(10), 69(1964).

804.31   Glausser, W.E., *Chem. Eng. Prog.*, **60**(10), 67(1964).

804.32   Snow, A.I., and W.S. Dickinson, *Chem. Eng. Prog.*, **60**(10), 64(1964).

804.33   Jamison, R.H., *Chem. Eng. Prog.*, **65**(3), 46(1969).

804.34 Formisono, F.A., *Chem. Eng.,* **76**(24), 108(Nov. 3, 1969).

804.35 Kister, H.Z., *Chem. Eng.,* **87**(26), 55(Dec. 29, 1980).

804.36 Troyan, J.E., *Chem. Eng.,* **68**(6), 147(Mar. 20, 1961).

804.37 Kister, H.Z., *Chem. Eng.,* **87**(15), 79(July 28, 1980).

804.38 Kister, H.Z., *Chem. Eng.,* **87**(10), 138(May 19, 1980).

804.39 Kister, H.Z., *Chem. Eng.,* **87**(18), 119(Sept. 8, 1980).

804.40 Thorogood, R.M., *I. Chem. E. Symp. Ser.,* **61**, 95(1981).

804.41 Fair, J.R., and R.L. Matthews, *Petrol Refiner,* **37**(4), 153(1958).

804.42 Sakata, M., *Chem. Eng. Prog.,* **62**(11), 98(1966).

804.43 Kister, H.Z., *Hydroc. Proc.,* **58**(2), 89(1979).

804.44 Barker, P.E., *Brit. Chem. Eng.,* **8**(5), 306(1963).

804.45 Davies, J.A., *Chem. Eng. Prog.,* **61**(9), 74(1965).

804.46 Grohse, E.W., R.F. McCartney, H.J. Hauer, J.A. Gerster, and A.P. Colburn, *Chem. Eng. Prog.,* **45**(12), 725(1949).

804.47 Kastanek, F., and G. Standard, *Separation Sci.,* **2**(4), 439(1967).

804.48 Lockett, M.J., C.T. Lim, and K.E. Porter, *Trans. Instn. Chem. Engrs.,* **51**, 61(1973).

804.49 Lockwood, D.C., and W.E. Glausser, *Petrol. Refiner,* **38**(9), 281(1959).

804.50 Loud, G.D., and R.C. Waggoner, *Ind. Eng. Chem. Proc. Des. Dev.,* **17**(2), 149(1978).

804.51 Love, F.S., *Chem. Eng. Prog.,* **71**(6), 61(1975).

804.52 Gagneux, A.L., and A.B. Hiser, *Petrol. Refiner,* **33**(3), 165(1954).

804.53 Buckley, P.S., R.K. Cox, and D.L. Rollins, *Chem. Eng. Prog.,* **71**(6), 83(1975).

804.54 Bolles, W.L., *Chem. Eng. Prog.,* **72**(9), 43(1976).

804.55 Klein, G.F., *Chem. Eng.,* **89**(9), 81(May 3, 1982).

804.56 Thorngren, J.T., *Ind. Eng. Chem. Proc. Des. Dev.,* **11**(3), 28(1972).

804.57 Glitsch, Inc., "Ballast Tray Design Manual," Bulletin No. 4900, 3rd Ed., Dallas, TX.

804.58 Koch Engineering Co., Inc., "Flexitray Design Manual," Bulletin 960-T, Wichita, KS.

804.59 Nutter Engineering Co., "Float Valve Design Manual," Tulsa, OK.

804.60 Lockett, M.J., and I.S. Ahmed, *Chem. Eng. Res. Des.,* **61**(2), 110(1983).

804.61 Kister, H.Z., *Chem. Eng.,* **88**(3), 107(Feb. 9, 1981).